Meteorological drought and flood scenarios over Kenya, East Africa

Ayugi Brian Odhiambo et al.

ELIVA PRESS

ELIVA PRESS

Ayugi Brian Odhiambo

Kenya's economy and that of the Great Horn of Africa rely on rain-fed agriculture. Unfortunately, the country is prone to hydrological extremes, mainly drought and floods. The occurrence of drought/flood is associated with not only the destruction of property but with loss of lives. The situation is worsening under the observed reduction in seasonal rainfall and amplified global warming. Characterizing the recent evolution of climate extreme scenarios across localized domains remains an imperative process to adapt tailor suit innovative solutions to drought risks and their impacts. The purpose of this book is aimed at demonstrating the recent changes in droughts/pluvial events over Kenya for planning purposes. The first chapter describes the existing literature on the past work documenting historical occurrences of drought/flood over localized domains or the whole country. Chapter two presents the approaches used to examine the current drought/flood scenarios by illustrating the trend, intensity, severity, and frequency based on the Standardized Evapotranspiration Precipitation Index (SPEI). Chapter three describe the results derived from careful analysis of climate extremes most pronounced over the study region. Lastly, possible conclusions and recommendations are presented in the last chapter. The book is designed as an upper-level undergraduate, graduate, and research text.

Authors: Ayugi Brian Odhiambo, Guirong Tan, Niu Ruoyun, Dong Zeyao, Moses Ojara, Lucia Mumo, Hassen Babaousmail, and Victor Ongoma.

Published: Eliva Press SRL
Address: MD-2060, bd.Cuza-Voda, 1/4, of. 21 Chişinău, Republica
Moldova
Email: info@elivapress.com
Website: www.elivapress.com

ISBN: 978-1-63648-032-9

METEOROLOGICAL DROUGHT AND FLOOD SCENARIOS OVER KENYA, EAST AFRICA

AYUGI BRIAN

Collaborative Innovation Center on Forecast and Evaluation of Meteorological Disasters/Key Laboratory of Meteorological Disaster, Ministry of Education, Nanjing University of Information Science and Technology, Nanjing 210044, China

ABSTRACT

This work examines drought and flood events over Kenya from 1981 to 2016 using Standardized Precipitation-Evapotranspiration Index (SPEI). Spatiotemporal analysis of dry and wet events is conducted for 3 and 12-months. Extreme drought incidences were observed in the years 1987, 2000, 2006, and 2009 for SPEI-3 whilst the SPEI-12 demonstrated the manifestation of drought during the years 2000 and 2006. SPEI shows that the wettest period, 1997 and 1998, coincide with the El Nino event for both time steps. SPEI -3 shows a reduction in moderate drought events while severe and extreme cases were on the increase tendencies towards the end of the twentieth century. Conversely, SPEI-12 depicts an overall increase in severe drought occurrence over the study location with observed intensity of -1.54 and cumulative frequency of 64 months during the study period. Wet events are on upward trend in the western and central highlands while the rest of the regions show an increase in dry events during the study period. Moreover, moderate dry/wet events predominate whilst extreme events occur least frequent across all grid cell. It is apparent that the study area experiences mild extreme dry events in both categories although moderately severe dry events dominate most parts of the study area. High intensity and frequency of drought is noted in SPEI-3 while least occurrences of extreme events are recorded in SPEI-12. Although drought event prevails across the study area, there is evidence of extreme flood conditions over the recent decades. These findings form a good basis for next step of research that will look at projection of droughts over the study area based on regional climate models.

Table of Contents

1. INTRODUCTION

Drought remains one of the most complex natural phenomena affecting the economy, environment and society at global, regional and local level (Parsons et al., 2019). For instance, occurrences of prolonged rainfall failure remarkably alter water resources, ecosystem balance, and have adverse impact on agriculture and urban livelihoods (Wilhite, 2000; Rohli et al., 2016). There is growing concern following the impacts of rapidly changing climate with projections pointing to an increase in extreme events (such as droughts and floods) that are expected to occur in across many regions (IPCC, 2014).

Consequently, with emphasis on drought, the focus of many researchers has been to infer from the intricate dynamics of drought and vulnerability impacts in a bid to establish mitigation measures (Sheffield et al., 2012; Huang et al., 2017; Wang et al., 2018). Despite the efforts, according to World Meteorological Organization (WMO) WMO (2010), there is still limited understanding of drought evolution, frequency, and severity of occurrence from one region to another. This is due to its 'creeping phenomenon' as compared to other natural disasters (Wilhite, 2000). For instance, drought varies by multiple dynamic dimensions including severity and duration making it difficult for scientists and policy makers to determine the exact timing of its inception or termination of either meteorological, agricultural or hydrological drought events (Wilhite, 1993; Labedzi, 2007; Mishra and Singh, 2010; WMO, 2016)

Numerous studies have reported an upsurge in drought events in many regions with noticeable increase over the recent decades, because of the ongoing global warming and decadal variability (Dai, 2011a; Sheffield et al., 2012; Dai, 2013a; IPCC, 2014b; Trenberth et al., 2014). To illustrate this, drought has affected many countries in Europe (Bradford, 2000; Hoerling et al., 2012; Spinoni et al., 2015), (North America Cook et al., 2007; Swalm et al., 2012; AghaKouchak et al., 2014), Asia (Liang et al., 2014; Cai et al., 2015; Sun et al., 2016), Australia (Chiew et al., 2014; Rahmat et al., 2015; Park et al., 2019), and Africa (Hulme, 1992; Dai, 2011a; Lyon and DeWitt, 2012). Most significantly, Africa, southern Europe, and eastern Australia have recorded an increase in drought events, mostly attributed to precipitation decrease linked with decadal fluctuations in the Pacific and western Indian Ocean (Dai, 2013b; Hua et al., 2016; Dai and Zhao, 2017).

East Africa (EA), mainly classified as an arid and semi-arid (ASAL) region despite falling within the tropics, continues to experience unprecedented records of drought events in comparison to other natural threats such as heat waves, torrents, cold surge, and cyclones (Lyon, 2014; Gebremeskel et al., 2019). Colossal records of economic losses and environmental degradation continue to be witnessed across many parts of the region (Guha-Sapir et al., 2004; Nicholson, 2014). For example, Kenya, Uganda, Somalia, and Ethiopia experienced severe drought event in 2010-2011 (Gebremeskel et al., 2019), with an estimated 10 million people acutely impacted (Nicholson, 2014). Furthermore, approximately 450,000 deaths were reported in Ethiopia during the 1984-1985 drought while Kenya witnessed a wide spread drought in 2005, affecting 2.5 million people in the northern region (Balint et al., 2013; Guha-Sapir et al., 2004). This trend is likely to increase with intensification of extreme climate events towards the end of the 21st century (Dai, 2011a; Rowell et al., 2015; Ongoma et al., 2018). Global predictions based on Palmer Drought Severity Index (PDSI) show that desiccation will become more severe and widespread over EA region with reduced precipitation and increased evaporation (Dai, 2013a).

Kenya has been witnessing an increase in severe and frequent famine events in the recent decades, exacerbated by the recent decline in March-May (MAM) seasonal rainfall (William and Funk, 2011; Funk et al., 2012; Ongoma and Chen, 2017; Ayugi et al., 2018). Numerous studies have been conducted to ascertain drought variabilities, trends and the respective impacts on agriculture, economy, water resources and environment over the study region region (Changwony et al., 2017; Karanja et al., 2017; Mutsotso et al., 2018; Polong et al., 2019). These researchers have employed various drought indices recommended by the (WMO, 2016). For instance, Mutsotso et al. (2018) investigated the drought occurrences in Kenya based on the combination of Standardized Precipitation-Evapotranspiration Index (SPEI) and Normalized Difference Vegetation Index (NDVI) on one-month basis and three-month basis and analyzed correlation between the two indices. On the other hand, Karanja et al. (2017) used Standardized Precipitation Index (SPI) to characterize seasonal and annual droughts in Laikipia west sub-county, Kenya from 1984 to 2014. The study focused on drought events occurring during the two rain seasons, namely March-May (MAM) and October-December (OND). Frank et al. (2017), employed Effective Drought Index (EDI), as an "accurate" index in drought assessment in Tana River basin in Kenya. In contrast, Zargar et al. (2011) reported contradictory results that EDI seemed to have weak imprecision in monitoring the inception, cessation and accumulated stress. Wambua et al. (2018) applied both SPI and EDI to delineate drought occurrences during 1980-2016 in the upper Tana River Basin, where nearly all agro-ecological zones of Kenya are located. Both indices demonstrated that the southeastern parts of the basin were more likely to experience severe droughts as compared to the northwestern parts.

The mentioned indices employed by various researchers over the study domain highlighted a glimpse of spatiotemporal variation and occurrence of historical dry/wet events from one region to another, without necessarily indicating the magnitude, severity, and duration of extreme events. Moreover, other studies on drought and flood evaluation reported a contrary occurrence of dryness/wetness events while some showed incoherence in spatial patterns of drought frequencies. Therefore, precise analysis of recent changes in drought and wet events in a complex subtropical domain is an important step in identifying mechanisms associated with these anomalous events in the era of changing climate, which remains a challenge.

Thus, the main objective of this study is to characterize drought and wet events based on intensity, severity, and frequency at each grid cell over the Kenya from 1981 to 2016, using widely accepted and used index, SPEI (Vicente-Serrano et al., 2010). The results are useful in accurate examination of the drought and wet events in the study locale, thereby helping hydrologist and farmers to take timely decisions. The findings of this work form a good basis for analysis and discussion of drought projections over the region based on improved Regional Climate Models (RCM) datasets. The remaining sections are organized as follows: Section 2 highlights characterization of the study area, data and methodology while results and discussions are given in section 3. Finally, conclusion and recommendation are presented in part 4.

2. CHAPTER TWO
2.1 Study area

Kenya is situated in East Africa. It is bound within longitude 34° E- 42° E and latitude 5° S - 5° N (**Fig. 1**). Adjoining nations include Uganda, Tanzania and Somalia. The economy of the country is predominately anchored on rain-fed agriculture (Mumo et al., 2018). Complex geomorphological features regulating local climate dominate different parts in the country. Highest altitude is in central highlands while low-lying regions characterized by ASAL ecosystems and climate occupies eastern, northwest and northeastern sides. Towards the south, lies the Indian Ocean coastline regulating local climate whereas the western sides of the study domain have large water basin of Lake Victoria, driving land-lake breezes (Camberlin, 2018).

The rainfall of the study locale is mostly bimodal with 'long rains' experienced during March to May (MAM) while 'short rains' occur during October to December (OND) (Ayugi et al., 2016, 2018, 2019; Ongoma and Chen, 2017; Mumo et al., 2019). Overall, a dry anomalous climate is experienced despite the region being situated along equatorial wet tropical belt. Numerous researches have examined the influencing factors regulating the rainfall variability (Kinuthia and Asnani, 1982; Indeje et al., 2000; Hasterath et al., 2004; Funk et al., 2013; Liebmann et al., 2014; Nicholson, 2014; Ogwang et al., 2015; Ayugi et al., 2018). The seasonal rainfall is mainly as an influenced by oscillation of Intertropical convergence zone (ITCZ) and local mesospheric factors (Indeje et al., 2000; Nicholson, 2017) whereas interannual variability is mainly influenced by global teleconnection dynamics (Hasterath, 2000; Pohl and Camberlin, 2011). Downward trend of long rains has been reported in many studies (Funk et al., 2008; Lyon and Dewitt, 2012; Yang et al., 2014; Ongoma and Chen 2017; Ayugi et al., 2018; Mumo et al., 2019) and an opposite trend in short rains (Ongoma and Chen, 2017). This is of great concern to farmers who have for long rely on long rains season for planting purposes due to changes in climatic patterns. Nicholson (2017) and Camberlin (2018) elaborate more on general circulation features of the study domain.

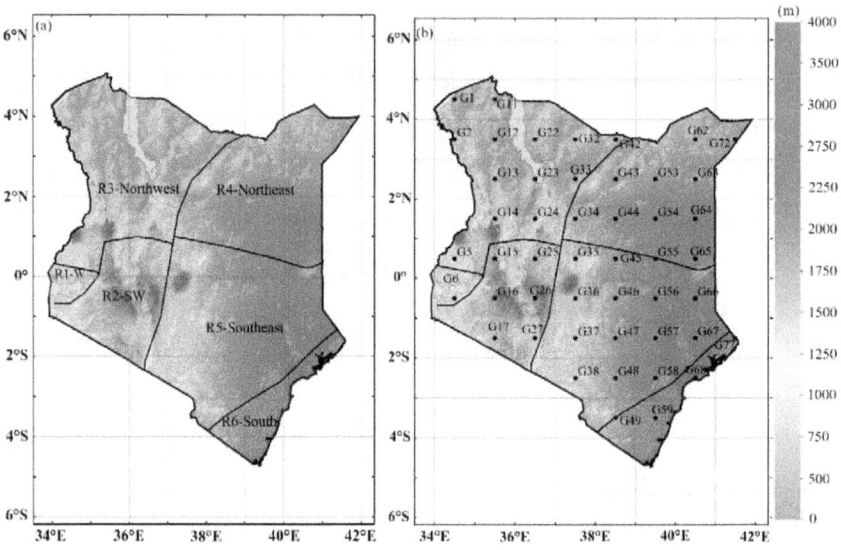

Figure 1. *Elevation of the study area with distinct homogeneous locations as delineated by (Indeje et al. Indeje et al., 2000) and the respective grid cells in each region.*

2.2 Data Descriptions

Comprehensive assessment of meteorological drought and wet events over a region involves use of several climatic datasets. This study utilized monthly maximum and minimum temperature datasets from Climatic Research Unit (CRU TS4.03; Harris et al., 2014) and monthly precipitation datasets obtained from Climate Hazard Group Infrared Precipitation with Station (CHIRPS.v2; Funk et al., 2015).

The CHIRPS datasets were acquired for the years 1981-2016. The datasets were produced following two steps: (i) pentad rainfall estimates, produced from cold cloud duration (CCD) based satellite data on regression model calibrated using TRMM; (ii) the stations are merged with CHIRP data to produce CHIRPS . This product has spatial resolution of 0.05 (~ 5.3 km) with a quasi-global coverage (50° S – 50° N, 180° E – 180°

W) and is existing from 1981 to present-day at pentad, decadal, and monthly temporal resolution (Funk et al., 2015).

The CHIRPS were recently evaluated by inferring their performance over the study domain (Ayugi et al., 2019a, 2019b) The CRU data with spatial resolution of ~ 50 km X 50 km was used in deriving the potential evapotranspiration (PET). In the present study, all datasets were extracted from all grid cell within the study domain (**Fig. 1**). This was derived from re-gridding of study area based on 1° x 1° spatial resolution in bid to achieve uniform grids for analysis since the gridded datasets were of varying resolutions. Analysis at each grid cell provides an insight to evolution of extreme events in region that is characterized by varying topographical features on finer horizontal resolution. This approach improves the representation of orographic features, such as elevation, land use and other surface features which might not otherwise be captured in major homogeneous regions (Polong et al., 2019).

2.3 Methods and Metrics
2.3.1 Standardized Precipitation Evapotranspiration
The SPEI is computed using precipitation and PET to delineate the phases of anomaly of dry and wet conditions by normalizing the alteration amongst water supply (precipitation) and demand (evapotranspiration). The SPEI and SPI (McKee et al., 1993). are almost similar except that SPEI includes PET and employs various schemes to derive PET. The SPEI built in R Program language version 3.4.2 (http://cran.rproject.org/web/packages/SPEI) was used to compute the SPEI. Vicente-Serrano et al. (2010) expounds more details on the mathematical equation for computing SPEI.

Comparable to the original SPI, a negative value indicates dry conditions, whilst positive value depicts wet condition (Gozzo et al., 2019). For instance, drought events are divided into four main categories, namely: extreme (SPEI ≤ -2.00), severe (-1.50> SPEI > -1.99), moderate (-1.00 > SPEI > -1.49), and mild (0 >SPEI > -0.99). Similarly, wet events are categorized as follows: extreme (SPEI $\leq$ +2.00), severe (+1.50 > SPEI > +1.99), moderate (+1.00 > SPEI > +1.49), and mild (0 > SPEI > +0.99). These values for SPEI, defines the characteristics of drought or wet condition in terms of severity, intensity, and duration of occurrence. In this study, the threshold for SPEI $\leq$ -1.0 was inferred to signify dry events whereas SPEI $\geq$ +1.0 represent wet events over the study domain. Similar threshold was employed in a study recent study that examined spatiotemporal evolution of drought in the Tana River Basin, Kenya (Polong et al., 2019).

This study defined the severity, intensity, and frequency for dry/wet event over the study domain as given in Equations 1 – 3;

i)   Severity is the cumulative sum of the index value based on the duration extent (Equation 1);

$$S = \sum_{i=1}^{Duration} Index \qquad (1)$$

ii)   Intensity of an event is the severity divided by the duration (Equation 2). Events that have shorter duration and higher severity will have large intensities.

$$I = \frac{Severity}{Duration} \qquad (2)$$

iii)   Frequency of occurrence (F_s) is defined in the Equation 3;

$$F_s = \frac{n_s}{N_s} x 100\% \qquad (3)$$

where n_s is the number of drought events (SPEI < -1.0), N_s is the total of the months for the study period, and s is a grid cell.

Furthermore, the duration of dryness/wetness situation is presented by the length of time (months) that the drought index is consecutively above or below a truncation value. The intensity, severity and frequency of extreme events define drought/wet episodes. The dominance of the dry/wet cases was examined for each grid cell and timescale and computed on the percentage of frequency of each incidence with reference to the total number of months. This approach was successfully employed in a recent study of drought evaluation along the major water basin in Kenya (Polong et al., 2019). The intention of employing this approach was to categorize regions that frequently experience concurrence of extreme and severe climatic cases at corresponding periods.

The SPEI values were calculated in two time scales: namely, SPEI-3 and SPEI -12. The SPEI-3 is derived by averaging the 3-month values, i.e. (March-May) within a year while SPEI-12 is from accumulated 12-month timescale. A timescale 3 month was chosen to denote drought/flood impacts on agriculture during the crop growing season (Hulme, 1996; Hayes et al., 1999; Balint et al., 2011). On the other hand, selection of a 12-month timescale aimed to reflect hydrological consequences of drought such as energy production services

2.3.3 Mann-Kendall test

The study employed Mann-Kendall (MK) test (Mann, 1945; Kendall, 1975) to detect the significance of the SPEI-3 and SPEI-12 analysis. The non-parametric feature of MK test allows it to confirm the existence of trend in any data against the null hypothesis of no trend. In addition, it does not require the sample to conform to any specific probability distribution since it works well even with insufficient or abnormal values. Significance of the trend was tested at 5% significance level. A Z-score exceeding the magnitude of critical values at 5% significance level denotes a monotonic trend of drought/flood events. The S value denotes variance, which is used to calculate significance of the trend. Numerous hydro-climate studies across various domains have employed MK tools for trend analysis (Araghi et al., 2016; Ongoma and Chen, 2017; Ayugi and Tan, 2019).

2.3.3 Empirical Cumulative Frequency

The Empirical Cumulative Frequency (ECDF) (Zambreski, 2016), was adopted to compare the performance of time series for SPEI and SPI over the study region. Empirical Cumulative Frequency $(F_N(t))$ is expressed as given in Equation 4;

$$F_N(t) = \frac{1}{N} \sum_{i=1}^{N} 1(x_i \leq t) \qquad (4)$$

where N is the number of months in observations and $(x_i \leq t)$ is the number of drought index values than the value of t. The concurrence performance of SPI and SPEI based on ECDF analysis over study domain is presented in **Figure 2**. The results show that the indices share similar frequency for all drought categories as highlighted in (Gozzo et al., 2016). For instance, the cumulative frequency for severe and extreme wet/dry events is well captured across the classifications. Besides, the robustness of SPEI is derived from its ability to combine the various aspects of the SPI with data on evapotranspiration, qualifying it further to a substantially accurate drought index. Many evaluative studies have ranked SPEI as the best index for drought assessment compared to other indices (McKee et al., 1993; Lorenzo-Lacruz et al., 2010). The PET employed in the present study is based on Hargreaves scheme that relies on any available time series datasets and has superior performance similar to that of Food and Agricultural Organization (FAO) criterion of Pen-Monteith (Allen et al., 1998). Comparative studies examining the suitability of different PET estimations over diverse domains concluded better performance of the PET derived from the Hargreaves equation with cautionary point regarding difference in few hundred-of-millimeter scale across different locations or characterized by unique land cover (Federer et al., 1996). Thus, the present study employed SPEI indices for historical trends synthesis of drought at each grid pixel over the study domain. SPI is recommended for drought/flood events analysis in situation where minimum (maximum) temperature datasets to aid in computing the evapotranspiration for SPEI are missing or unavailable.

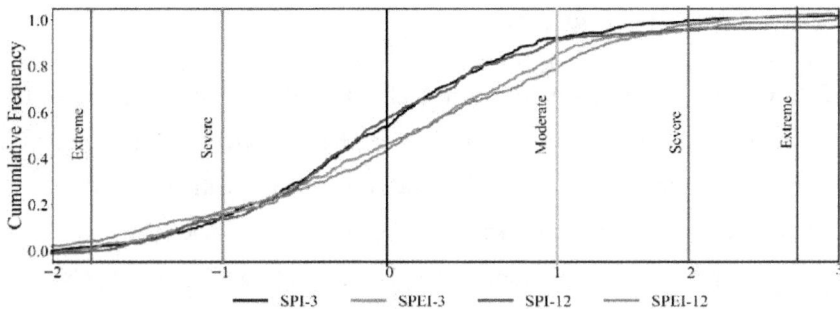

Figure 2. *The empirical cumulative frequency of the monthly SPI and SPEI for Kenya region. Vertical lines represent dry spell (red) and wet spell (green) threshold classified in drought indices presented in Gozzo et al. [92].*

3. CHAPTER THREE: RESULTS
3.1 Temporal patterns and frequency incidences of dry/wet events

Figure 3 provides an overview of historical analysis for SPEI 3- and 12 for the years 1981 to 2016 over Kenya. In addition, the Mann-Kendall test statistics for SPEI-3 and SPEI-12 analysis over study domain is presented in **Table 1**. The evaluation of drought and wet events was conducted for moderate, severe and extreme frequencies (Vicente-Serrano et al., 2010; Begueria et al., 2014). From the SPEI-3 results (**Figure 3a**), it can be seen that the study domain experiences moderate to severe and moderate to extreme drought cases towards the end of the twentieth century. SPEI-3 for both wet and dry events show an increasing trend at 5 % significant level (**Table 1**). The results agree with observed abrupt shift in rainfall tendency that occurred in the late 1990s from wet years to an almost continuous period with well-below average rainfall over the study domain (Funk et al., 2008; Williams and Funk, 2011;Lyon and DeWitt, 2012) This trend has been consistent in the three decades following the early 1980s and a series of very dry years prevailed around 2010 (Ayugi et al., 2018). The extreme drought incidences were observed during the years 1987, 2000, 2006, and 2009 for SPEI-3. The listed years coincide with atmospheric circulation changes related to SSTs variations that influence the regional rainfall patterns (Hua et al., 2016). Further, the observed changes in drought characteristic for SPEI-3 event indicate moderate intensity phenomenon at -1.43, although the severity recorded is more intense with noted value of -111.5 over the duration of 78 months (**Table 2**). It is apparent from results presented that SPEI-3 exhibit greater temporal frequency of occurrence of wet and dry cases during the study duration. This could be explained by the fact that SPEI-3 represent an average of months in which region receives maximum rainfall amount characterized by high variability from one region to another.

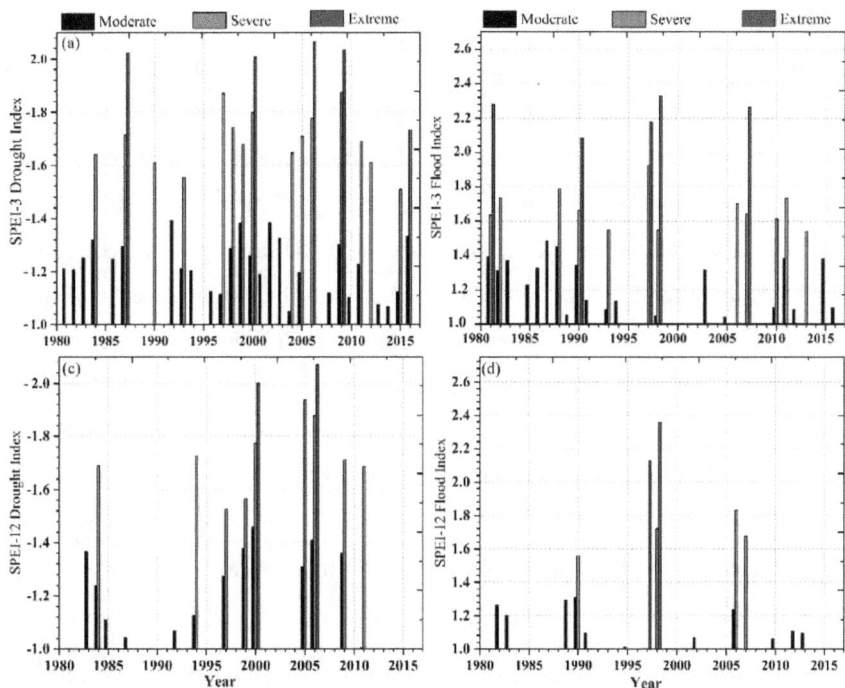

Figure 3. *Evolution of the mean SPEI for 3- and 12-month timescale for moderate, severe and extreme drought and flood over Kenya showing the variation in the duration, severity and intensity of dry and wet events, 1981-2016.*

Trend analysis	SPEI-test			
	SPEI-3 dry	SPEI-3 wet	SPEI-12 dry	SPEI-12 wet
S	3150	1130	2080	-1170
Z	13.12	11.38	11.77	-11.77
p	<0.0001	<0.0001	<0.0001	<0.0001
Alpha	0.05	0.05	0.05	0.05
significance	Significant Increasing	Significant Increasing	Significant Increasing	Significant Decreasing

Table 1. *Summary of Mann-Kendall test statistic for SPEI-3 and SPEI-12 over Kenya at 5 % significant level. Negative (positive) Z-values indicate significantly or insignificant decreasing (increasing) trend.*

The results for SPEI-12 show stability in the frequency of incidences over the study area during the period 1981-2016. This demonstrates that SPEI at elongated timescales respond gradually and consistently to deviations in climatic variables indicating strong durations of frequent occurrences of anomalous events over the years. Subsequently, the longer timescales are most appropriate for the revealing of incidences of signature events over the region whereas shorter intervals demonstrate suitability for detecting frequent seasonal and inter-annual variations. Further analysis of drought severity (**Figure 3c**) show an overall severe drought occurrence over the study location with observed intensity of -1.54 and cumulative frequency of 64 months during the study period (**Table 2**). The extreme drought incidences were observed during the years 2000 and 2006. Significant test at 5 % significance reveal a decreasing trends for SPI-12 wet years while dry years show opposite increasing tendencies (**Table 1**). Overall decreasing patterns in wet events is mainly influenced by the observed decrease in rainfall over the study area during the recent decades (Lyon, 2014, Ongoma et al., 2017, Ayugi et al., 2018; Mumo et al., 2018).

Comparison of the two indices show that SPEI-3 is characterized by severe drought occurrence while long-term drought (SPEI-12) show reduction in the extreme events over the study area. Meanwhile, the wetness episodes for both SPEI-3/12 demonstrate severe occurrence with intensity of $SPEI \geq 1.5$ in all intervals. The wettest period between 1997 and 1998 during the El Nino event is well captured in both time steps, depicting a robust performance of SPEI in capturing the underlying mechanisms of wet conditions. A comparison of the two results reveals an overall moderate dry conditions occurrence while more intense wet events over the short duration of existence are experienced.

	SPEI	Duration	Severity	Intensity
Dry	3	78	-111.15	-1.43
	12	64	-98.70	-1.54
Wet	3	61	94.79	1.55
	12	61	93.31	1.53

Table 2 *The duration, severity, and intensity occurrence of the major dry ($SPEI \leq -1$) and wet $SPEI \geq 1$) events over Kenya during 1981 to 2016.*

Further evaluation for SPEI were conducted over six homogeneous climatic zones as delineated by Indeje et al. (2000). The SPEI values for each region was identified by averaging the values at grid cell as presented in **Figure 1**. The regions are as follows: R1-western sides; R2-Southwest; R3-Northwest; R4-Northeast; R5-Southeast; and R6-South coastal area. **Figures 4 and 5** demonstrate the linear trend for dry and wet events for different time scales across the six regions during the period of 1981 to 2016. It is worth noting that in both timescales, the R1, and R2 depict increasing trend in wet events while the rest of the regions show increase in dry events during the study period. R1 and R2 is characterized by high elevations while R3, R4, and R5 is mostly occupied by the bare lands of arid and semi-arid climate. These regions experience below-normal rainfall and high temperature, resulting to high evapotranspiration as compared to R1 and R2 which is characterized by dense vegetation cover and raised water table. Moreover, the high terrains in R1 and R2 produce lee rain shadows and block the passing of rain bearing disturbances in other regions (Ogwang et al., 2014). In-depth analysis at each grid cell along the homogenous zones was conducted based on duration, severity and magnitude of occurrences of some significant anomalous incidences. **Table 3** highlights the evolution of dry/wet events for some noteworthy cases over the study region. The dry/wet years highlighted concurred with similar years as noted over the whole area average (**Figure 3**).

The results from analyses of frequency of wet and dry events for SPEI-3 for all grids pixels across the study domain demonstrate that moderate events predominate while extreme events occur less frequently across all grid cells during the study period. It is further noted that variations of wet/dry events occur across different grid cells from one-time scale to another. This agrees with previous studies that observed similar variations of anomalous climate events across different parts of the study domain (Awange et al., 2007; Karanja et al., 2017; Changwony et al., 2017; Frank et al., 2017). For instance, the wettest event for SPEI-3 was recorded in grid cell 58 during May 1981 and the duration for occurrence lasting for 87 months. Likewise, the prolonged duration of dry event for SPEI-3 was observed in pixel 36 and 49 lasting for 83 months. Regarding the severity of below and above normal events over the study domain, **Table 3** gives locations which had experienced these abnormal climatic cases. The most severe dry event for SPEI-3 was remarked in grid cell 36. More notably, the severe wet event was experienced over similar grid cell 36 with a magnitude of 125.95. This shows that the region experiences both extremes as compared to other regions over the study domain, a feature worth further investigation.

The findings for SPEI-12 based on individual grid cell for the analysis of wet and dry events frequency demonstrate that moderate events prevail while extreme events occur less frequently across all grid cell during the study period. The wettest event for SPEI-12 was experienced March 1998 in grid cell 34 and 55 with the duration of occurrence persisting for 97 months. Drought analysis demonstrated prolonged duration of dry event in grid cell 7 for 82 months. Concerning the severity of below and above normal events over the study domain, **Table 3** show that most severe dry event for SPEI-12 ensued in pixel 62. On the contrary, the severe wet event was noted over grid cell 17 and 63 with a magnitude of 119. Overall, moderate intensity of both wet and dry events for SPE-3 and 12 was experienced across the study domain during the study duration except for grid cell 55 that recorded high intensity of SPEI $\geq$ 1.5. This agrees with previous studies conducted over various parts of the study domain or based on different indices (Mwangi et al., 2014; Nicholson et al., 2014; Mutsotso et al., 2016).

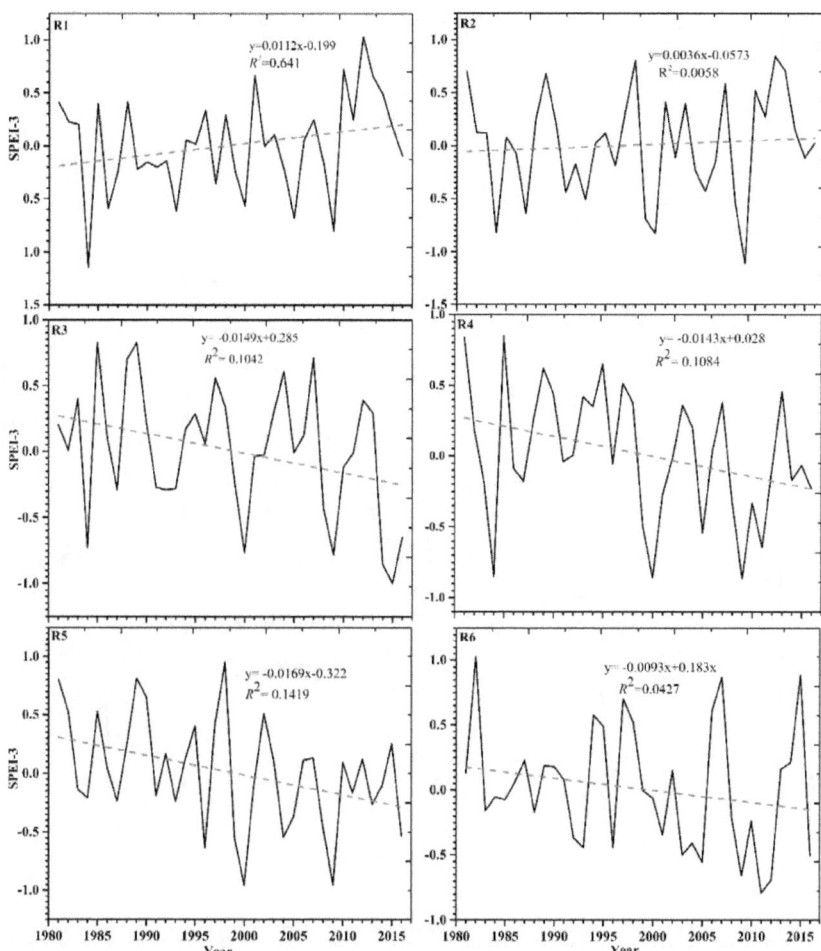

Figure 4. *Linear trends of dry and wet events for SPEI-3 over six distinct climatic zones as presented in* **Figure 1**

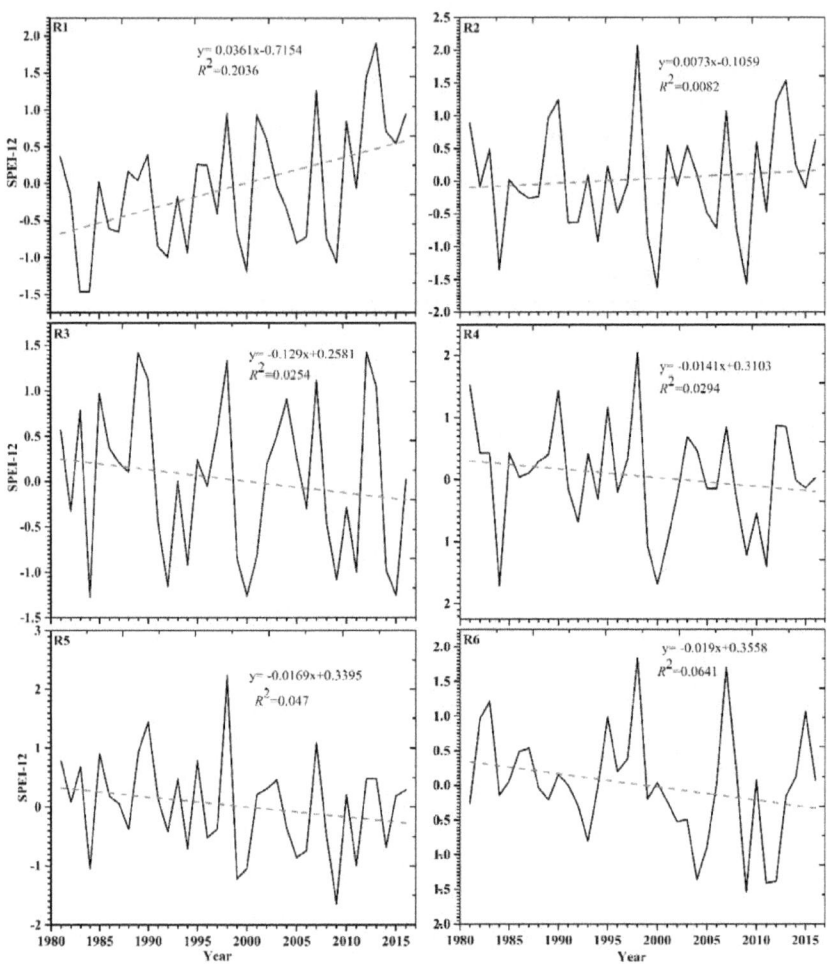

Figure 5. *Linear trends of dry and wet events for SPEI-12 over six distinct climatic zones as presented in* **Figure 1**

Grid	Duration	Severity	Intensity	Grid	Duration	Severity	Intensity
Dry event for SPEI-3				Wet event for SPEI-3			
4	80	-110.74	-1.38	4	81	115.69	1.42
15	81	-109.76	-1.35	26	80	117.26	1.46
16	81	-109.74	-1.35	44	85	121.71	1.43
27	82	-114.28	-1.39	48	81	117.61	1.45
36	83	-114.81	-1.38	72	87	125.92	1.44
49	83	-112.69	-1.35	73	80	119.58	1.49
67	80	-107.9	-1.34	76	80	116.05	1.45
Dry event for SPEI-12				Wet event for SPEI-12			
7	82	-112.69	-1.37	7	81	111.65	1.37
13	80	-112.28	-1.40	27	80	116.22	1.45
17	78	-110.48	-1.41	32	85	119.88	1.41
36	76	-107.69	-1.41	38	81	115.89	1.43
48	78	-108.77	-1.39	47	80	113.5	1.42
55	79	-108.21	-1.36	53	97	97.59	1.54
62	78	-113.83	-1.45	58	81	115.83	1.43
63	77	-112.19	-1.45	77	87	119.4	1.37

Table 3. The duration, severity and intensity of occurrence of some of the major dry and wet events (SPEI $\leq$ -1 SPEI $\geq$ +1) over each grid cell in Kenya, 1981- 2016

3.3 Spatial patterns of SPEI in the study area

The spatial pattern of frequency of severe and extreme dry (wet) cases for the SPEI-3 and 12 months period are presented in **Figures 6 and 7,** respectively. From the analysis introduced in **Figure 6**, it is apparent that the study area experiences mild extreme dry events in both categories whilst moderately severe dry events dominate the most parts of the domain. For instance, the frequency of severe dry events for SPEI-3 varied from 3.25 in grid cell 72 to 6.51 in pixel 57, situated in northeastern region. On the other hand, extreme dry events depicted uniform distribution over the study locale with low incidences during SPEI-3. The percentage occurrence ranged from 0.23 in grid cell 27 and 36 to 3.72 in grid cell 53, along northeastern area. Moreover, maximum frequency recorded in pixel 63 with percentage value of 1.86. Overall, high intensity and frequency of drought are noted during SPEI-3.

Analyses for severe and extreme wet events are presented in **Figure 7.** The wet events were mainly found in the western parts, extending towards southern sides and partly central areas during the severe wet events for SPEI-3 while extreme wetness covered most parts of the study domain. The frequency of severe wet events for SPEI-3 varied from 3.25 in grid cell 59 and 50 to 7.9 in pixel 5, situated in western region. On the other hand, maximum frequency was recorded in pixel 59 with percentage value of 3.25 whereas minimum wet event observed over grid 6 along Lake Victoria, with recorded value of 0.46.

Meanwhile, assessment for SPEI-12 revealed occurrence of maximum severe dry event observed in grid cell 42 (8.6) characterized by ASAL ecosystems while least severe dry scenario recorded over in grid 77 (2.09) along coastal region. Over ASAL areas, anomalous soil moisture content restraints dehydration as well as little vegetation cover results to depressed transpiration rates, ensuing in low mean latent heat flux as compared to over humid lands [6]. Extreme dry events depict bimodal distribution over the study locale with least occurrences of extreme events recorded during the study duration.

The severity and extreme wet events are presented in **Figure 7.** The wet events were mainly noted in the western parts, extending towards southern sides and partly central areas during the severe wet events while strong wetness is observed over northeastern during the SPEI-12 timescales. The frequency of severe wet events for SPEI-12 was observed in grid cell 16 (9.06) while minimum severe wet event observed over most grids namely, 64, 65, and 43 (0.23), situated along the northeastern sides. Meanwhile, extreme wet events had percentage occurrence ranging from 0.23 over grid cell 23 to 4.18 in grid cell 34.

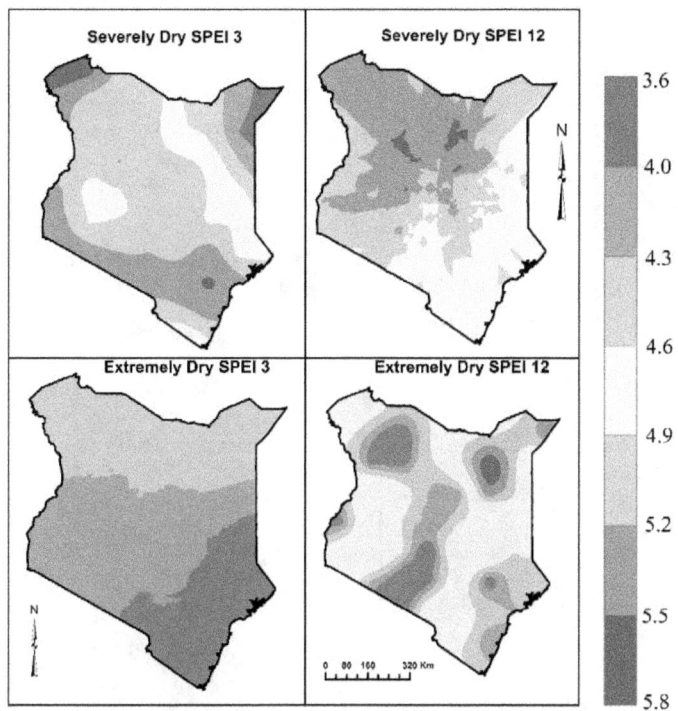

Figure 6. *Frequency of severe (top) and extreme (bottom) dry events computed for the SPEI-3 and SPEI-12 month respectively over Kenya, 1981-2016.*

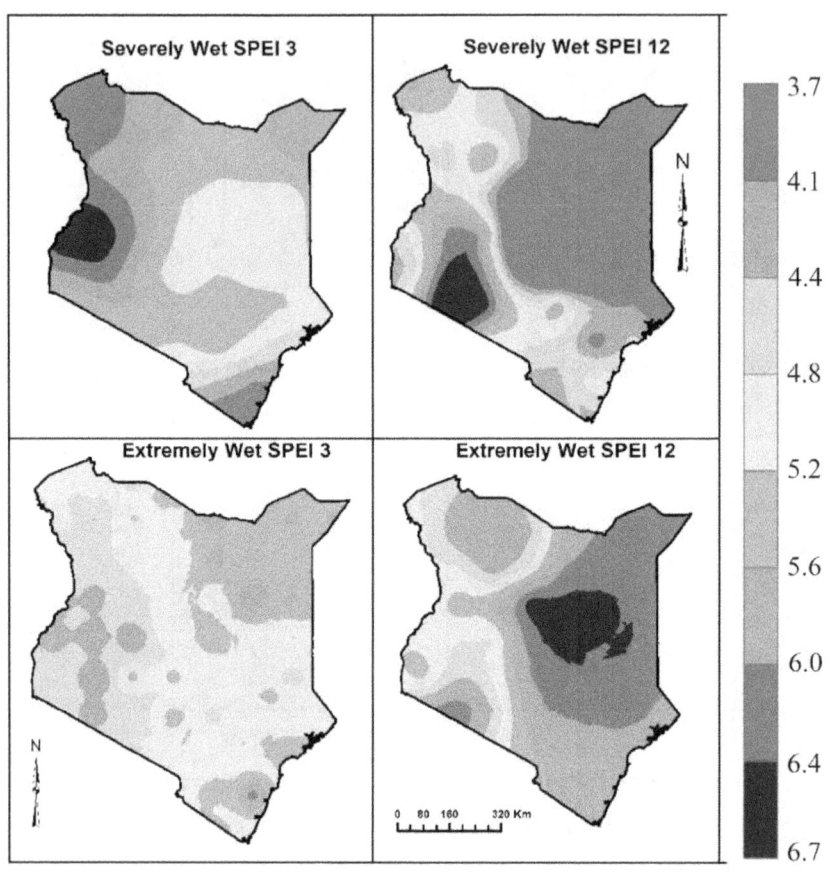

Figure 7. *Frequency of severe (top) and extreme (bottom) wet events computed for the SPEI-3 and SPEI-12 month respectively over Kenya, 1981-2016*

DISCUSSION AND CONCLUSION

4.1 Discussions

Drought occurrence is a stochastic natural phenomenon that is mainly influenced from the changes in climatic variables, namely precipitation and temperature. The variability of both variables on historical perspective have been observed a number of existing studies with sharp increase being noted, most significantly in temperature towards the end of twentieth century and beginning of twenty first century era (Ongoma et al., 2018). It is worth remarking that the observed trends towards the end of 21st century are mainly as a result of GHG-induced changes and changes caused by internal variability, e.g by ENSO and the Inter-decadal Pacific Oscillation (IPO) (Gu et al., 2013; Lyon, 2014, Dong et al., 2015; Dai, 2016). This is equally echoed in a recent Intergovernmental Panel on Climate Change (IPCC) report that stated unequivocal warming from 1950s across the globe, as a result of anthropogenic induced global warming (IPCC, 2014). The observed variability in climatic variables is likely to have a profound impact in drought/flood mechanism over study domain, which is influenced by ratio of precipitation and potential evapotranspiration (Hulme, 1996; Feng and Fu, 2013; Lyon, 2014). For instance, increase in surface air temperature towards the end of the 21st century will likely significantly influence the PET level which characterizes the evaporative demand of the atmosphere. Other factors include the low humidity and abundant solar radiation which remains a signature feature over the study domain (Ji et al., 2015).

As a consequence of global warming, there is apprehension that increased temperature which is linked to evapotranspiration may lead to increase in drought incidences and severity across many regions (Dai, 2011, 2013; Liebmann et al., 2014). The study domain has been experiencing rapid increase of trends in extreme events characterized by drought and wet scenarios towards end of 20th century and beginning of 21st century (**Table 1**). While drought event has prevailed, there are extreme flood conditions with devastating consequences, which are equally witnessed (Nicholson, 2014). Evidently, the severity and intensity of drought/flood, along with abrupt deviations between the extremes continues to pose a threat to the livelihoods of people hence, the need for continuous evaluation of drought and flood occurrences over the study domain remains paramount.

The results of this present study which examined meteorological drought and wet scenarios over Kenya using a robust index of SPEI during study duration (**Figure 3**), points to decreasing patterns in moderate wetting occurrences towards the end of twentieth century which also in harmony with past studies (Funk et al., 2008; Williams and Funk, 2011; Tierney et al., 2015). Moreover, the impact of drought is shown to vary from low-lying region to humid vegetative areas (**Figure 4-5**), predominantly due to surface and atmospheric interaction dynamics. For example, the changes in wet events is mostly associated with heightened heating Sea Surface Temperatures (SST) of Indian Ocean, which alters the Walker circulation anomalies contributing to drying trends (Hua, et al., 2016). On the other hand, the recent drying tendency over the study domain (**Table 3**) could be because of changes in the tropical SSTs variations over Indo-Pacific (Williams and Funk, 2011; Liebmann et al., 2014). This may lead to substantial increase in regional aridity and drought areas (Dai, 2011; 2013). In addition, the large-scale atmospheric circulation changes associated with a weaker West African monsoon may likely to have contributed (Lyon, 2012; Hua et al., 2016). Consequently, this could affect negatively on the area's economy that entirely relies on season-based farming for livelihoods and sustainability (Funk et al., 2008; Mumo et al., 2018)

This study thus highlights key features of drought and wet events, which remain to be major climatic extremes occurrence affecting people and property (Zhang et al., 2019; Dilling et al., 2019). Understanding historical complexities and dynamics of drought and flood events build a momentum to conduct further studies on future evolution of these extreme events over the study area with view of recommending appropriate and timely policies to avert damages and loss of life. This study reveals occurrences of mild extreme dry events whilst moderately severe dry events dominate over most parts of the domain (**Figure 6 and 7**). High intensity and frequency of drought are noted in SPEI-3 whereas least occurrences of extreme events are recorded in SPEI-12.

4.2 Conclusion

Drought remains to be one of the most complex natural phenomena affecting the economy, environment and society at global, regional and local level. The present study examines drought and wet events by characterizing the trends, intensity, severity and frequencies based on widely accepted indices of SPEI over Kenya, East Africa from 1981 to 2016. The spatial and temporal evolution of dry and wet events is captured by both 3- and 12- month SPEI. Extreme drought incidences were observed during the years 1984, 1987, 2000, 2006, and 2009 for SPEI-3 whilst for SPEI-12 drought manifested during the years 2000 and 2006. The wettest period was noted in 1997-1998 attributed to a strong El Nino event. This shows how well SPEI performs in capturing the underlying mechanisms of dry/wet conditions.

SPEI -3 shows an occurrence of moderate to severe and moderate to extreme drought cases towards the end of the twentieth century whilst SPEI-12 depicts an overall increase in severe drought occurrence over the study location with observed intensity of -1.54 and cumulative frequency of 64 months during the study period. Spatial patterns show that western and central highlands depict increasing trend in wet events while the rest of the regions show increase in dry events during the study period. Moreover, moderate dry/wet events are dominant while extreme events occur least frequent across all grid cells during the study period.

It is apparent that the during the study duration, the region experiences mild extreme dry events in both categories whilst moderately to severe dry events dominate over most parts of the domain. High intensity and frequency of drought are noted in SPEI-3 whilst least occurrences of extreme events are recorded in SPEI-12. Whereas drought event has prevailed, there are extreme flood conditions with possible devastating consequences equally witnessed.

ACKNOWLEDGEMENT : The authors acknowledge Nanjing University of Information Science and Technology (NUIST) for providing favorable environment and infrastructural needs for conducting research. Special appreciation to all data centers for availing data to use for evaluation studies.

LITERATURE

Allen, R., Pereira, L.D., Raes, and M. Smith., 1998. Crop evapotranspiration guidelines for computing crop water requirements, FAO Irrig. Drain. Pap. 56, Food and Agric. Organ. of the U. N., Rome

Araghi A., Mohammad MB., Jan A., 2016. Detection of trends in days with extreme temperatures in Iran from 1961 to 2010. Theor. Appl. Climatol. 125, 213–225. http://doi.org/10.1007/s00704-015-1499-6

Awange, J.L., Aluoch, J., Ogallo, L.A., Omulo, M., Omondi, P., 2007. Frequency and severity of drought in the Lake Victoria region (Kenya) and its effects on food security. Clim Res.33, 135–142. https://doi.org/10.3354/cr033135

Ayugi, B.O., Wen, W., Chepkemoi, D., 2016. Analysis of Spatial and Temporal Patterns of Rainfall Variations over Kenya. J. Environ. Earth Sci.6(11).

Ayugi B.O, Tan G, Ongoma V, Mafuru KB., 2018a. Circulations Associated with Variations in Boreal Spring Rainfall over Kenya. Earth Syst Environ. 2, 421–434. https://doi.org/10.1007/s41748-018-0074-6

Ayugi B.O, Tan, G 2018b. Recent trends of surface air temperatures over Kenya from 1971 to 2010. Meteorol. Atmos. Phys., https://doi.org/10.1007/s00703-018-0644-z

Ayugi, B., Tan, G., Ullah, W., Boiyo, R., Ongoma, V., 2019a. Inter-comparison of remotely sensed precipitation datasets over Kenya during 1998–2016. Atmos. Res. 225, 96-109.https://doi.org/10.1016/j.atmosres.2019.03.032

Ayugi, B., Tan, G., Gnitou, GT., Ojara, M., Ongoma, V., 2019b. Historical evaluations and simulations of precipitation over Eastern Africa from Rossby Centre Regional Climate Model. Atmos. Res. 232. https://dio.org/10.1016.jatmosres.2019.104705

AghaKouchak, A., L. Cheng., M. Omid, and F. Alireza., 2014. Global warming and changes in risk of concurrent climate extremes: Insights from the 2014 California drought, Geophys. Res. Lett. 41, 8847-8852.

Balint, Z., Mutua, F. M., Muchiri, P., 2011. Drought monitoring with the combined drought index. *FAO-Swalim, Nairobi, Kenya*, 3-25.

Balint, Z., Mutua, F., Muchiri, P., Omuto, C.T., 2013. Monitoring drought with the combined drought index in Kenya. : A Natural Outlook-Geo-Environmental Resources and Hazards. Earth surf. Proc. Land., 341-356. https://dio.org/10.1016/b978-0-444-59559-1.00023-2

Bradford, R.B., 2000. Drought events in Europe. In: Vogt, J.V., Somma, F. (Eds.), Drought and Drought Mitigation in Europe. Kluwer, Dordrecht, pp. 7–20.

Beguería, S., Vicentc-Scrrano, S. M., Reig, F., Latorre, B.,2014. Standardized precipitation evapotranspiration index (SPEI) revisited: parameter fitting, evapotranspiration models, tools, datasets and drought monitoring. Int. J. Climatol. 34, 3001-3023.

Camberlin, P., 2018. Climate of Eastern Africa (Vol. 1).Oxford Research Encyclopedia of Climate Science. https://doi.org/10.1093/acrefore/9780190228620.013.512

Cai, Q., Y. Liu, H. Liu, and J. Ren., 2015. Reconstruction of drought variability in North China and its association with sea surface temperature in the joining area of Asia and Indian-Pacific Ocean, *Palaeogeography, Palaeoclimatology, Palaeoecology*, 417, 554-560

Changwony, C., Sichangi, A. W., Murimi Ngigi, M., 2017. Using GIS and Remote Sensing in Assessment of Water Scarcity in Nakuru County, Kenya. *Advances in Remote Sensing*, 06, 88–102. https://doi.org/10.4236/ars.2017.61007

Chiew, F.H.S., Potter, N.J., Vaze, J., Petheram, C., Zhang, L., Teng, J., Post, D. A., 2014. Observed hydrologic non-stationarity in far south-eastern Australia: Implications for modelling and prediction. Stoch Environ Res Risk Assess, 28, 3–15. https://doi.org/10.1007/s00477-013-0755-5

Cook, E. R., Seager, R., Cane, M. A., Stahle, D. W., 2007. North American drought: Reconstructions, causes, and consequences. Earth Sci. Rev, 81, 93–134. https://doi.org/10.1016/j.earscirev.2006.12.002

Dai, A., 2011a. Characteristics and trends in various forms of the Palmer Drought Severity Index during 1900-2008. J Geophys Res. 116. https://doi.org/10.1029/2010JD015541

Dai, A., 2011b. Drought under global warming: A review. WIRES Clim Change,2, 45–65. https://doi.org/10.1002/wcc.81

Dai, A., 2013a. Increasing drought under global warming in observations and models. Nature Clim Change, 3, 52–58. https://doi.org/10.1038/nclimate1633

Dai A., 2013b. The influence of the inter-decadal Pacific Oscillation on U.S. precipitation during 1923–2010. Clim. Dyn, 41: 633–646. DOI https://doi.org/10.1007/s00382-012-1446-5

Dai, A., 2016. Future Warming Patterns Linked to Today's Climate Variability. Sci Rep,6, 6–11. https://doi.org/10.1038/srep19110

Dai A, Zhao T., 2017. Uncertainties in historical changes and future projections of drought. Part I: estimates of historical drought changes. Climatic Change.144:519–33. https://doi.org/10.1007/s10584-016-1705-2

Dong, B., Dai, A., 2015. The influence of the Interdecadal Pacific Oscillation on Temperature and Precipitation over the Globe. Clim Dynamics. 45, 2667–2681. https://doi.org/10.1007/s00382-015-2500-x

Federer, C.A., 1996. Intercomparison of methods for calculating potential evaporation in regional and global water balance models. Water Resour. Res, 32, 2315–2321.

Feng, S., Fu, Q., 2013. Expansion of global drylands under a warming climate. Atmos Chem Phys, 13, 10081–10094. https://doi.org/10.5194/acp-13-10081-2013

Frank, A., Armenski, T., Gocic, M., Popov, S., Popovic, L., Trajkovic, S., 2017. Influence of mathematical and physical background of drought indices on their complementarity and drought recognition ability. Atmos. Res. 194, 268-280. https://doi.org/10.1016/j.atmosres.2017.05.006

Funk, C., Dettinger, M.D., Michaelsen, J.C, Verdin, J.P., Brown, M.E., Barlow, M., Hoell, A., 2008. Warming of the Indian Ocean threatens eastern and southern African food security but could be mitigated by agricultural development. P. Natl. Acad. Sci. USA,

105, 11081–11086. https://doi:10.1073/pnas.0708196105

Funk, C.C., 2012. Exceptional warming in the western Pacific-Indian Ocean warm pool has contributed to more frequent droughts in eastern Africa. Bull. Amer. Meteor. Soc.,93: 1041–1067. doi:10.1175/BAMS-D-12-00021.1

Funk, C., Husak, G., Michaelsen, J., Shukla, S., Hoell, A., Lyon, B., Hoeling, M.P.,Liebmann B, Zhang, T., Verdin, J.,Galu, G., Eilerts, G., Rowland, J., 2013. Attribution of 2012 and 2003 – 12 Rainfall Deficits in Eastern Kenya and Southern Somalia. Bull. Amer. Meteor. Soc, 94(9), S45-S48.

Funk, C., Peterson, P., Landsfeld, M., Pedreros, D., Verdin, J., Shukla, S., Michaelsen, J., 2015. The climate hazards infrared precipitation with stations - A new environmental record for monitoring extremes. Scientific Data, 2, 1–21. doi.org/10.1038/sdata.2015.66.

Gebremeskel, G., Tang, Q., Sun, S., Huang, Z., Zhang, X., Liu, X., 2019. Droughts in East Africa: Causes, impacts and resilience. Earth-Sci Rev, 193, 146–161. https://doi.org/10.1016/j.earscirev.2019.04.015

Gu, G., Adler, R.F., 2013. Interdecadal variability/long-term changes in global precipitation patterns during the past three decades: Global warming and/or pacific decadal variability? Clim Dyn, 40, 3009–3022. https://doi.org/10.1007/s00382-012-1443-8

Guha-Sapir, D., Hargitt, D., Hoyois, P., 2004. Thirty years of natural disasters 1974-2003: The numbers. Presses univ. de Louvain.

Harris, I., Jones, P.D., Osborn, T.J., Lister, D.H., 2014. Updated high-resolution grids of monthly climatic observations – the CRU TS3.10 Dataset. Int. J. Climatol.,34: 623–642. https://doi.org/10.1002/joc.3711

Hastenrath, S., 2000. Zonal circulations over the equatorial Indian Ocean. J. Climate 13(15), 2746–2756. https://doi.org/10.1175/1520-0442(2000)013<2746:ZCOTEI>2.0.CO;2

Hastenrath, S., Lamb, P. J., 2004. Climate dynamics of atmosphere and ocean in the equatorial zone: A synthesis. Int. J. of Climatol. 24, 1601-1612. https://doi.org/10.1002/joc.1086

Hastenrath, S., Polzin, D., Mutai, C., 2011. Circulation mechanisms of kenya rainfall anomalies. J. Climate24(2), 404–412. https://doi.org/10.1175/2010JCLI3599.1

Hayes MJ, Svoboda MD, Wilhite DA, Vanyarkho OV., 1999. Monitoring the 1996 drought using the Standardized Precipitation Index. Bull Am Meteor Soc 80:429–438

Hoerling, M., Eischeid, J., Perlwitz, J., Quan, X., Zhang, T., Pegion, P., 2012. On the increased frequency of Mediterranean drought. J. Clim. 25, 2146–2161. https://doi.10.1175/JCLI-D-11-00296.1

Hua, W., Zhou, L., Chen, H., Nicholson, S.E., Raghavendra, A., Jiang, Y., 2016. Possible causes of the Central Equatorial African long-term drought. Environ. Res. Lett. 11 (12), 124002.

Huang, J., Guan, X., Ji, F., 2012. Enhanced cold-season warming in semi-arid regions. Atmos Chem Phys.12, 5391–5398. https://doi.org/10.5194/acp-12-5391-2012

Huang, J., Li, Y., Fu, C., Chen, F., Fu, Q., Dai, A., Wang, G., 2017. Dryland climate change: Recent progress and challenges. Rev. Geophys, 55, 719–778. https://doi.org/10.1002/2016RG000550

Hulme, M., 1992. Rainfall changes in Africa: 1931–1960 to 1961–1990. Int. J. Climatol., 12, 685-699. https://doi.org/10.1002/joc.3370120703

Hulme, M., 1996. Climatic change within the period of meteorological records. In: Adams WM, Goudie AS, Orme AR (eds) The physical geography of Africa. Oxford University Press, Oxford, p 88–102

IPCC (2014b) Climate Change 2014: Synthesis Report. Contribution of Working Groups I, II and III to the Fifth Assessment Report of the Intergovernmental Panel on Climate Change [Core Writing Team, Pachauri RK, Meyer LA (eds)]. IPCC, Geneva, Switzerland, 151 pp.

IPCC (2014a) Summary for policymakers, Climate Change 2014: Impacts, Adaptation, and Vulnerability, Contribution of Working Group II to the Fifth Assessment Report of the Intergovernmental Panel on Climate Change [Field CB, Barros VR, Dokken DJ, Mach KJ, Mastrandrea MD, Bilir TE, Chatterjee M, Ebi KL, Estrada YO, Genova RC, Girma B, Kissel ES, Levy AN, MacCracken S, Mastrandrea PR, White LL (eds.)]. Cambridge University Press, Cambridge, United Kingdom and New York, NY, USA, pp. 1–32.

Indeje, M., Semazzi, F.H.M., 2000. Relationships between QBO in the Lower Equatorial Stratospheric Zonal Winds and East African Seasonal Rainfall. Meteorol. Atmos. Phys. 73, 227–244. https://doi.org/10.1007/s007030050075.

Ji, M., Huang, J., Xie, Y., Liu, J., 2015. Comparison of dryland climate change in observations and CMIP5 simulations. Adv. Atmos Sci, 32, 1565-1574. https://doi.org/10.1007/s00376-015-4267-8

Karanja, A., Ondimu, K., Recha, C., 2017. Analysis of Temporal Drought Characteristic Using SPI Drought Index Based on Rainfall Data in Laikipia West Sub-County, Kenya. OALib, 04, 1–11. https://doi.org/10.4236/oalib.1103765

Kendall, M.G., 1975. Rank Correlation Methods, 4th ed. Griffin, London, 202 pp.

Kinuthia, J.H., Asnani, G.C., 1982. A newly found jet in North Kenya (Turkana Channel). Mon. Weather Rev.110 (11), 1722–1728. https://doi:10.1175/1520-0493(1982)110<1722%3AANFJIN>2.0.CO%3B

Labedzi., 2007. Estimation of Local Drougtht Frequency in central Poland using the Standardized Precipitation Index SPI. Irrig. and Drain, 56, 67–77. https://doi.org/10.1002/ird.285

Liang, L., Zhao, S.H., Qin, Z.H., Ke-Xun, H. E., Chong, C., Luo, Y.X., Zhou, X.D., 2014. Drought change trend using MODIS TVDI and its relationship with climate factors in China from 2001 to 2010. J. of Integrative Agric.13, 1501–1508. https://doi.org/10.1016/S2095-3119(14)60813-3

Liebmann, B, Hoerling M.P., Funk, C., Bladé, I., Dole, R.M., Allured, D., Quan, X., Pegion P., Eischeid, J.K., 2014. Understanding Recent Eastern Horn of Africa Rainfall Variability and Change. J. Climate,27: 8630–8645. doi:10.1175/JCLI-D-13-00714.1

Lorenzo-lacruz, J., Vicente-serrano, S.M., López-moreno, J.I, Beguería, S.,García-ruiz, J.M., 2010. The impact of droughts and water management on various hydrological systems in the headwaters of the TagusRiver (central Spain). J Hydrol 386, 13–26. https://doi.org/10.1016/j.jhydrol.2010.01.001

Lyon, B., Dewitt, D.G., 2012. A recent and abrupt decline in the East African long rains. Geophys. Res. Lett.,39, L02702. https://doi:10.1029/2011GL050337

Lyon, B., 2014. Seasonal drought in the Greater Horn of Africa and its recent increase during the March–May long rains. J. Clim. 27 (21), 7953–7975.

Manatsa, D., Mukwada, G., Siziba, E., Chinyanganya, T., 2010. Analysis of multidimensional aspects of agricultural droughts in Zimbabwe using the Standardized Precipitation Index (SPI). Theor Appl Climatol, 102, 287–305. https://doi.org/10.1007/s00704-010-0262-2

Mann, H.B., 1945. Nonparametric tests against trend. Econometrica 13, 245-259

McKee, T.B, Doesken, N.J, Kleist, J., 1995. Drought monitoring with multiple time scales. In: 9th conference on applied climatology, Dallas, TX. Amer. Meteorol. Soc., pp 233–236

McKee, E., McKee, T.B., 2012. World Meteorlogical Organization, 2012: Standardized Precipitation Index User Guide (M. Svoboda, M. Hayes and D. Wood). (WMO-No. 1990), Geneva.

Mishra, A.K., Singh, V.P., 2010. A review of drought concepts. J. Hydrol, 391, 202–216. https://doi.org/10.1016/j.jhydrol.2010.07.012

Mutsotso, R.B., Sichangi, A.W., Makokha, G.O., 2018. Spatio-Temporal Drought Characterization in Kenya from 1987 to 2016. Adv. Remote Sens., 07, 125–143. https://doi.org/10.4236/ars.2018.72009

Mumo, L., Yu, J., Ayugi, B., 2019. Evaluation of spatiotemporal variability of rainfall over Kenya from 1979 to 2017. J. Atmospheric Sol-Terr Phys, 194, 105097. https://doi.org/10.1016/j.jastp.2019.105097

Mwangi, E., Wetterhall, F., Dutra, E., Di Giuseppe, F., Pappenberger, F., 2014. Forecasting droughts in East Africa. Hydrol. Earth Syst. Sci, 18, 611–620. https://doi.org/10.5194/hess-18-611-2014

Nicholson, S.E., 2014. A detailed look at the recent drought situation in the Greater Horn of Africa. J. Arid Environ. 103, 71-79. https://doi.org/10.1016/jaridenv.2013.12.003

Nicholson, S.E., 2014. The predictability of rainfall over the Greater Horn of Africa. Part I. Prediction of seasonal rainfall. J. Hydometeor, 15,1011-1027. doi.org/10.1175/JHM-D-13-062.1.

Nicholson, S.E., 2017. Climate and climatic variability of rainfall over eastern Africa. Rev. Geophys. 55, 590–635. https://doi.org/10.1002/2016RG000544

Ogwang, B.A, Ongoma, V., Li, X., Ogou, F.K., 2015. Influence of Mascarene High and Indian Ocean Dipole on East African Extreme Weather Events. Geogr. Pannonica19: 64–72. https://doi.105937/Geopan15020640

Ongoma, V., Chen, H., 2017. Temporal and spatial variability of temperature and precipitation over East Africa from 1951 to 2010. Meteorol. Atmos. Phys., 29: 131–144.doi:10.1007/s00703-016-0462-0

Ongoma, V., Chen, H., Gao, C., 2018a. Projected Change in Mean Rainfall and Temperature over East Africa based on CMIP5 Models. Int. J. Climatol. 38: 1375–1392 https://doi.org/10.1002/joc.5252

Ongoma, V., Chen, H., Gao, C., Nyongesa, A.M., Polong, F., 2018b.Future changes in Climate Extreme over Equatorial East Africa based on CMIP5 multimodel ensemble. Nat Haz, 90, 901–920.doi.org/10.1007/s11069-017-3079-9

Ongoma V, Tan G, OgwangBA, Ngarukiyimana JP (2015) Diagnosis of Seasonal Rainfall Variability over East Africa: A case study of 2010-2011 Drought over Kenya. *Pakistan Journal of Meteorology 11(22):* 13–21

Parsons, D. J., Rey, D., Tanguy, M., Holman, I. P., 2019. Regional variations in the link between drought indices and reported agricultural impacts of drought. Agric Sytm, 173, 119–129. https://doi.org/10.1016/j.agsy.2019.02.015

Polong, F., Chen, H., Sun, S., Ongoma, V., 2019. Temporal and spatial evolution of the standard precipitation evapotranspiration index (SPEI) in the Tana River Basin , Kenya. Theor Appl Climatol. https://doi.org/10.1007/s00704-019-02858-0

Pohl, B., Crétat, J., Camberlin, P., 2011. Testing WRF capability in simulating the atmospheric water cycle over Equatorial East Africa. Clim. Dyn, 37, 1357–1379. doi.org/10.1007/s00382-011-1024-2.

Rahmat, S. N., Jayasuriya, N., Bhuiyan, M., 2015. Development of drought severity-duration-frequency curves in Victoria, Australia. Australasian Journal of Water Resources, 19, 31-42. https://10.7158/13241583.2015.11465454

Rohli, R. V., Bushra, N., Lam, N.S. N., Zou, L., Mihunov, V., Reams, M. A., Argote, J.E., 2016. Drought indices as drought predictors in the south-central USA. Nat Hazards, 83, 1567–1582. https://doi.org/10.1007/s11069-016-2376-z

Rowell, D.P., Booth, B.B.B., Nicholson, S.E., Good, P., 2015. Reconciling past and future rainfall trends over East Africa. J. Clim. 28 (24), 9768–9788. https://doi.org/10.1175/JCLI-D-15-0140.1

Sen, P.K., 1968. Estimates of the Regression Coefficient based on Kendall's Tau. J. Amer. Stat. Assoc.,63: 1379–1389. https://doi:10.2307/2285891

Sheffield, J., Wood, E.F., Roderick, M.L., 2012. Little change in global drought over the past 60 years. Nature, 491, 435–438. https://doi.org/10.1038/nature11575

Sneyers, R., 1990. On the Statistical Analysis of a Series of Observations. Tech. Note 143, WMO-No. 415, 192.

Spinoni, J., Naumann, G., Vogt, J. V., Barbosa, P., 2015. The biggest drought events in Europe from 1950 to 2012. J. Hydrol: Regional Studies, 3, 509–524. https://doi.org/10.1016/j.ejrh.2015.01.001

Schwalm, C.R., C.A., Williams, K., Schaefer, D., Baldocchi, T.A., Black, A.H., Goldstein, B.E., Law, W.C., Oechel, K.T., Paw, R.L., Scott., 2012. Reduction in carbon uptake during turn of the century drought in western North America, Nature Geoscience, 5, 551-556. https://doi.org/1038/ngeo1529

Sun, S., Chen, H., Wang, G., Li, J., Mu, M., Yan, G., Zhu, S., 2016. Shift in potential evapotranspiration and its implications for dryness/wetness over Southwest China. J. Geophysical Research : Atmospheres. J. Geophys Res : Atmos, 4211–4232. https://doi.org/10.1002/2016JD025276

Tierney, J.E., Ummenhofer, C.C., deMenocal, P.B., 2015. Past and future rainfall in the Horn of Africa. Sci. Adv., 1(9), e1500682. https://doi:10.1126/sciadv.1500682

Trenberth, K.E., Dai, A., van der Schrier, G., Jones P.D., Barichivich J., Briffa K.R., 2014. Global warming and changes in drought. Nature Climate Change.;4:17–22. https://doi.org/10.1038/nclimate20167

Vicente-Serrano, S.M., S. Beguería., J. López-Moreno., 2010. A multiscalar drought index sensitive to global warming: the standardized precipitation evapotranspiration index, J. Climate, 23,1696-1718.https://doi.org/10.1175/2009JCLI2909.1

Vicente-Serrano, S.M., Beguería, S., Lorenzo-Lacruz, J., Camarero, J.J., López, Moreno, J.I., Azorin-Molina, C., 2012. Performance of drought indices for ecological, agricultural, and hydrological applications. Earth Interact. 16, 1–27. https://doi.org/10.1175/2012EI000434.1

Wambua, R.M., Benedict, M.M., James, M.R., 2018. Detection of Spatial, Temporal and Trend of Meteorological Drought Using Standardized Precipitation Index (SPI) and Effective Drought Index (EDI) in the Upper Tana River Basin, Kenya. Open Journal of Modern Hydrology, 08: 83–100. https://doi.org/10.4236/ojmh.2018.83007

Wang, G., 2005. Agricultural drought in a future climate: results from 15 global climate models participating in the IPCC 4th assessment. Clim dyn, 25, 739-753.https://doi.org/10.1007/s00382-005-0057-9

Wang, L., Yuan, X., Xie, Z., Wu, P., Li, Y., 2016. Increasing flash droughts over China during the recent global warming hiatus. Sci. rep, 6, 30571.https://doi.org/10.1038/srep30571

Wang, G., Gong, T., Lu, J., Lou, D., Hagan, D. F. T., Chen, T., 2018. On the long-term changes of drought over China (1948–2012) from different methods of potential evapotranspiration estimations. Int. J. Climatol, 38, 2954–2966. https://doi.org/10.1002/joc.5475

Wilhite, D.A., 2000. Chapter 1 Drought as a Natural Hazard: Concepts and Definitions. *Drought: A Global Assessment*, *1*, 3–18.

Williams, A.P., Funk, C., 2011. A westward extension of the warm pool leads to a westward extension of the Walker circulation, drying eastern Africa. Clim. Dyn.,37,2417–2435. https://doi:10.1007/s00382-010-0984-y

WMO., 2010. Experts Recommend Agricultural Drought Indices for improved understanding of food production conditions. Press Release No. 887, Geneva/Murcia, 8 June 2010.

World Meteorological Organization (WMO) and Global Water Partnership (GWP), 2016: Handbook of Drought Indicators and Indices (M. Svoboda and B.A. Fuchs). Integrated Drought Management Programme (IDMP), Integrated Drought Management Tools and Guidelines Series 2. Geneva.Wilhite, D.A., 1993. Understanding the phenomenon of drought. Hydrol. Rev. 12, 136–148.

World Bank., 2012. Doing business in the East African economies. IFC/World Bank Rep., 116

Yang, W., Seager, R., Cane, M.A., Lyon, B., 2014. The East African Long Rains in Observations and Models. J. Climate,27,7185–7202. https://doi.org/10.1175/JCLI-D-13-00447.1

Zargar, A., Sadiq, R., Naser, B., Khan, F.I., 2011. A review of drought indices. Environ Rev., 19, 333-349. https://doi.org/10.1139/a11-013

Publisher: Eliva Press SRL

Email: info@elivapress.com

Eliva Press is an independent publishing house established for the publication and dissemination of academic works all over the world. Company provides high quality and professional service for all of our authors.

Our Services:
Free of charge, open-minded, eco-friendly, innovational.

-Free standard publishing services (manuscript review, step-by-step book preparation, publication, distribution, and marketing).
-No financial risk. The author is not obliged to pay any hidden fees for publication.
-Editors. Dedicated editors will assist step by step through the projects.
-Money paid to the author for every book sold. Up to 50% royalties guaranteed.
-ISBN (International Standard Book Number). We assign a unique ISBN to every Eliva Press book.
-Digital archive storage. Books will be available online for a long time. We don't need to have a stock of our titles. No unsold copies. Eliva Press uses environment friendly print on demand technology that limits the needs of publishing business. We care about environment and share these principles with our customers.
-Cover design. Cover art is designed by a professional designer.
-Worldwide distribution. We continue expanding our distribution channels to make sure that all readers have access to our books.

www.elivapress.com

www.ingramcontent.com/pod-product-compliance
Lightning Source LLC
Chambersburg PA
CBHW051257170526
45165CB00004B/1752

* 9 7 8 1 6 3 6 4 8 0 3 2 9 *